Wow, That's Woolly!

ALPACAS

Gareth Stevens
PUBLISHING

Beth Gottlieb

Please visit our website, www.garethstevens.com. For a free color catalog of all our high-quality books, call toll free 1-800-542-2595 or fax 1-877-542-2596.

Library of Congress Cataloging-in-Publication Data

Names: Gottlieb, Beth, author.

Title: Alpacas / Beth Gottlieb.

Description: New York : Gareth Stevens Publishing, [2023] | Series: Wow, that's woolly! | Includes index.

Identifiers: LCCN 2021054489 (print) | LCCN 2021054490 (ebook) | ISBN 9781538279328 (set) | ISBN 9781538279335 (library binding) | ISBN 9781538279311 (paperback) | ISBN 9781538279342 (ebook)

Subjects: LCSH: Alpaca—Juvenile literature. | Alpaca (Textile)—Juvenile literature.

Classification: LCC SF401.A4 G68 2023 (print) | LCC SF401.A4 (ebook) | DDC 636.2/966—dc23/eng/20211201

LC record available at https://lccn.loc.gov/2021054489

LC ebook record available at https://lccn.loc.gov/2021054490

Published in 2023 by
Gareth Stevens Publishing
29 E. 21st Street
New York, NY 10010

Designer: Olivia Lo Sardo
Editor: Kristen Nelson

Portions of this work were originally authored by Alonso Garcia and published as *Alpacas*. All new material this edition authored by Beth Gottlieb.

Photo credits: Cover, pp. 1, 10, 12–13 murbansky/Shutterstock.com; Icons KozyrevaElena/Shutterstock.com; p. 5 Rita_Kochmarjova/Shutterstock.com; p. 6 javarman/Shutterstock.com; p. 7 lulu and isabelle/Shutterstock.com; p. 9 Solar760Lfoto/Shutterstock.com; p. 9 (map) Good_Stock/Shutterstock.com; p. 11 Nadia Kompan/Shutterstock.com; p.14 Delpixel/Shutterstock.com; p. 15 Christian Vinces/Shutterstock.com; p. 17 J. Lekauicius/Shutterstock.com; p. 19 Kira Volkov/Shutterstock.com; p. 20 SakuraImageInc/Shutterstock.com; p. 21 terekhov igor/Shutterstock.com

Printed in the United States of America

CPSIA compliance information: Batch #CSGS23: For further information, contact Gareth Stevens, New York, New York, at 1-800-542-2595.

CONTENTS

Words in the glossary appear in **bold** type the first time they are used in the text.

FRIENDLY AND FLEECY

Some of the warmest, softest **material** in the world is made from an animal that's small and quite woolly: the alpaca. Alpacas are covered in fleece, or wool, that comes in many colors. People have been using it to make clothing and other goods for thousands of years!

Alpacas are friendly, smart animals. They have been **domesticated** for many years, and they are comfortable with people. Alpacas can make good pets, but they are most often kept for their fleece.

PULL BACK THE WOOL

One way to tell alpacas apart from llamas is by their size. Llamas are about twice the size of alpacas!

Alpacas are part of the Camelidae family. They're related to llamas, vicuñas, and guanacos.

GET THE LOOK

There are two breeds, or kinds, of alpacas: suri and huacaya (wah-KAY-uh). Suri alpacas have long hair that hangs straight down. Huacaya alpacas have fluffy fleece that sticks out.

Alpacas come in about 20 different colors. They include white, black, brown, and many shades in between. Adult alpacas are about 3 feet (1 m) tall at the shoulder. They weigh between 100 and 175 pounds (45 and 79 kg). Alpacas can live for up to 20 years in **captivity**.

huacaya alpaca ▶

PULL BACK THE WOOL

Male alpacas are called machos and females are called hembras.

Huacaya alpacas are much more common than suri alpacas. Of the millions of alpacas in the world, less than 10 percent are suris.

suri alpaca

ALPACAS OF THE ANDES

Alpacas' natural **habitat** is in the Andes Mountains of South America. They have many **adaptations** that allow them to live in the harsh weather of the Andes.

Alpacas' thick fleece keeps them warm and dry during the cold, wet winters in the Andes. They have padded feet that allow them to walk on low-growing mountain grass without harming it as they **graze**. Alpacas have a long neck, which helps them see predators from far away. Their long legs help them run away from predators.

PULL BACK THE WOOL

Alpacas spit when they're upset!

Today, domesticated alpacas live on farms all over the world.

HERD LIFE

Alpacas live in groups called herds. They are very social animals that rely on the other alpacas around them. To talk to one another, alpacas hum! They may also scream or snort.

Alpacas are herbivores, which means they eat only plants. They mostly eat grass and hay, but they may also eat bark and wood. Alpacas spend a lot of their time chewing their food while laying down. But sometimes they "dance" together! This is called pronking.

PULL BACK THE WOOL

Herds of alpacas may also include goats, sheep, or even llamas.

Alpacas are commonly calm animals, so pronking can be quite a sight!

BABY ALPACAS

Alpacas are mammals, or **warm-blooded** animals that give birth to live young that drink milk made by their mothers.

Alpaca mothers give birth about 11 months after **mating**. The baby is called a cria. Alpacas commonly have only one baby at a time. At birth, a cria can weigh from 12 to 20 pounds (5.5 to 9 kg). It is able to stand and walk soon after it is born. When a cria is old enough to eat solid food, it's called a tui (TOO-ee).

PULL BACK THE WOOL

Male alpacas looking for a mate may "sing" to a female!

Crias are most often born between May and September.

ALPACA FARMS

People have been working with alpacas for thousands of years. The ancient people of South America domesticated alpacas to use them for their meat and fleece.

Today, many farms keep alpacas. They are very useful livestock and easy to care for. Some farmers keep alpacas as guard animals for sheep and chickens. Alpacas are tall enough to spot danger from a distance and can chase away foxes and dogs. Today, alpacas are most often kept by people for their fleece.

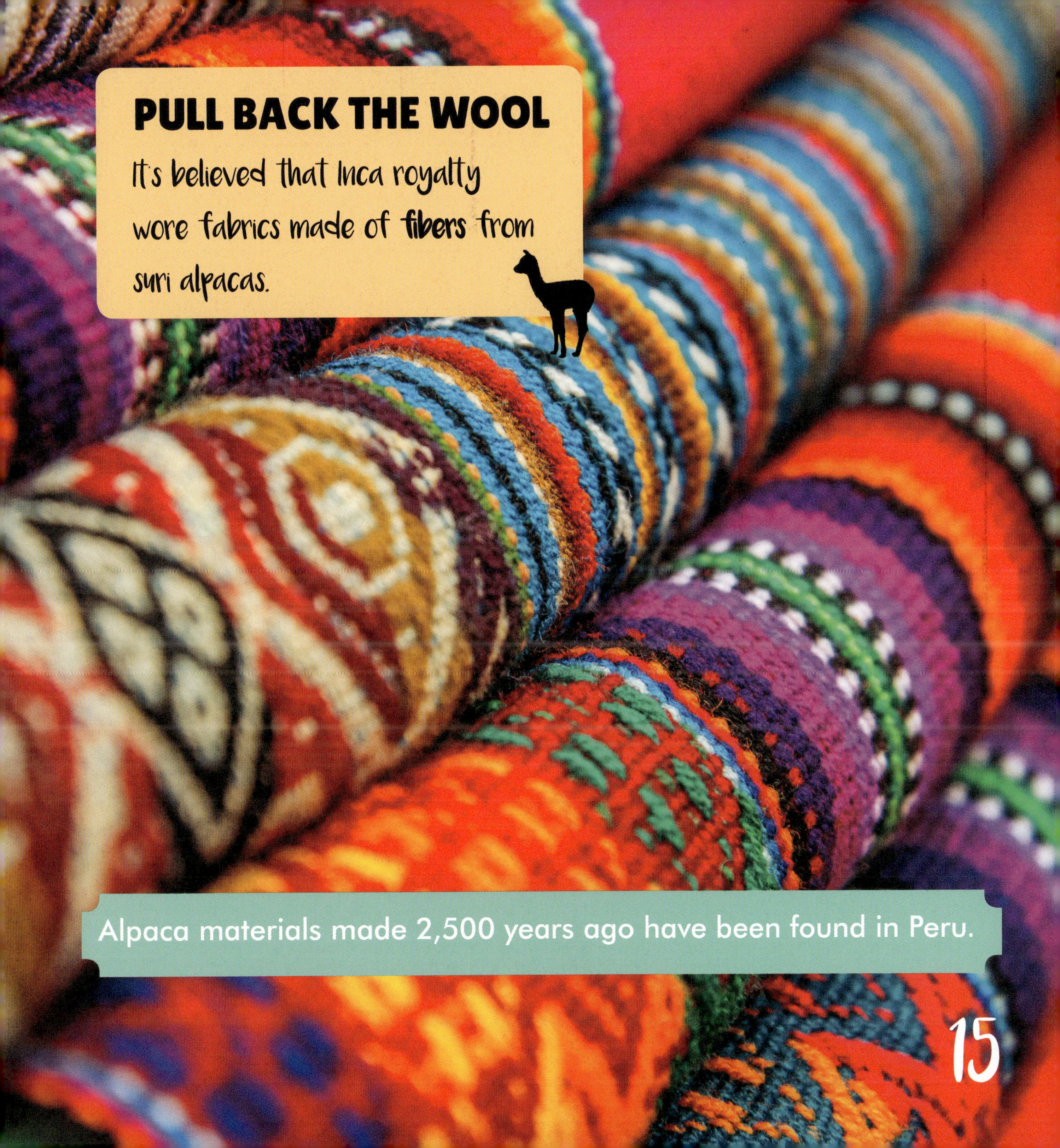

PULL BACK THE WOOL

It's believed that Inca royalty wore fabrics made of **fibers** from suri alpacas.

Alpaca materials made 2,500 years ago have been found in Peru.

STRONG AND SOFT

Alpaca fleece is one of the strongest natural fibers in the world. It's also warmer, softer, and more **water-resistant** than sheep's wool. The fiber has a sheen, or shiny appearance. Fineness and crimp are two important features of alpaca fiber. Fineness has to do with the thinness of the fiber, and crimp has to do with its shape.

Unlike many other fibers, alpaca fiber is **hollow**, which allows it to hold in more heat. It's also hypoallergenic, which means it's unlikely to cause an **allergic reaction**.

PULL BACK THE WOOL

Because it's less common, suri fleece and the material that's made from it are the most highly valued of alpaca products.

Alpaca fiber is worth so much money that selling it can cover the cost of keeping the animals.

SHEAR, SPIN, AND WEAVE

Farmers usually shear, or cut the fleece of, their alpacas during the spring so the animals don't overheat during the warmer months. Each alpaca can produce 5 to 10 pounds (2 to 4.5 kg) of fleece. After the fleece is cleaned, it's sorted based on color, length, and quality.

Spinning is the process by which raw alpaca fiber is made into yarn. Once the fiber is spun into yarn, it can be dyed. Yarn is then made into fabric through a process called weaving.

PULL BACK THE WOOL

Weaving is commonly done with a tool called a loom. The loom interlaces pieces of yarn to create a tight-knit piece of fabric.

Fleece from the back and belly is cut first because it's the most valuable. This is called blanket fleece.

WEAR ALPACA

Sweaters are one of the most common goods made from alpaca fleece. Outerwear like hats, gloves, and scarves are also made from alpaca fleece. Alpaca fleece is made into beautiful rugs and blankets as well.

Would you like to meet an alpaca? These woolly animals can be found on farms that often allow visitors. Some may even let you watch the alpacas being sheared a couple times a year. Alpacas may also be found at a zoo near you!

Goods made from alpaca fleece can cost a lot of money.

GLOSSARY

adaptation: A change in a type of animal that makes it better able to live in its surroundings.

allergic reaction: When a person's body responds badly to things in its surroundings that are usually harmless, such as dust, pollen, or insect stings.

captivity: The state of living in a place and not in the wild.

domesticate: To breed and raise for use by people.

fiber: A thin thread of natural or human-made material that can be used to make cloth.

graze: To feed on grass.

habitat: The natural home for plants, animals, and other living things.

hollow: Having nothing inside.

mate: To come together to produce babies. Also, one of the two animals that come together to produce babies.

material: Matter from which something is made.

warm-blooded: Able to keep the body at a steady temperature no matter what the outside temperature is.

water-resistant: Not allowing water to pass through easily.

FOR MORE INFORMATION

Books

Bové, Jennifer. *Ranger Rick: I Wish I Was a Llama.* New York, NY: HarperCollins, 2020.

Leaf, Christina. *Llama or Alpaca?* Minneapolis, MN: Bellwether Media, 2020.

Websites

Alpaca: The Maryland Zoo

www.marylandzoo.org/animal/alpaca/

Check out more fun facts about alpacas.

Fun Alpaca and Llama Facts for Kids

worldoftravelswithkids.com/fun-alpaca-and-llama-facts-for-kids/

Learn to tell llamas and alpacas apart with ease and much more about these woolly animals.

Publisher's note to educators and parents: Our editors have carefully reviewed these websites to ensure that they are suitable for students. Many websites change frequently, however, and we cannot guarantee that a site's future contents will continue to meet our high standards of quality and educational value. Be advised that students should be closely supervised whenever they access the internet.

INDEX